Wildland Fires
Florida - 1998

Investigated by: J. Gordon Routley

This is Report 126 of the Major Fires Investigation Project conducted by Varley-Campbell and Associates, Inc./TriData Corporation under contract EME-97-CO-0506 to the United States Fire Administration, Federal Emergency Management Agency.

Department of Homeland Security
United States Fire Administration
National Fire Data Center

U.S. Fire Administration Fire Investigations Program

The U.S. Fire Administration develops reports on selected major fires throughout the country. The fires usually involve multiple deaths or a large loss of property. But the primary criterion for deciding to do a report is whether it will result in significant "lessons learned." In some cases these lessons bring to light new knowledge about fire--the effect of building construction or contents, human behavior in fire, etc. In other cases, the lessons are not new but are serious enough to highlight once again, with yet another fire tragedy report. In some cases, special reports are developed to discuss events, drills, or new technologies which are of interest to the fire service.

The reports are sent to fire magazines and are distributed at National and Regional fire meetings. The International Association of Fire Chiefs assists the USFA in disseminating the findings throughout the fire service. On a continuing basis the reports are available on request from the USFA; announcements of their availability are published widely in fire journals and newsletters.

This body of work provides detailed information on the nature of the fire problem for policymakers who must decide on allocations of resources between fire and other pressing problems, and within the fire service to improve codes and code enforcement, training, public fire education, building technology, and other related areas.

The Fire Administration, which has no regulatory authority, sends an experienced fire investigator into a community after a major incident only after having conferred with the local fire authorities to insure that the assistance and presence of the USFA would be supportive and would in no way interfere with any review of the incident they are themselves conducting. The intent is not to arrive during the event or even immediately after, but rather after the dust settles, so that a complete and objective review of all the important aspects of the incident can be made. Local authorities review the USFA's report while it is in draft. The USFA investigator or team is available to local authorities should they wish to request technical assistance for their own investigation.

This report and its recommendations was developed by USFA staff and by Varley-Campbell and Associations, Incorporated (Miami and Chicago), its staff and consultants, who are under contract to assist the Fire Administration in carrying out the Fire Reports Program.

The United States Fire Administration greatly appreciates the cooperation received from the Florida Department of Forestry, Bunnell, Volusia County, Ormond Beach, St. John County and Brevard County Fire Department.

For additional copies of this report write to the United States Fire Administration, 16825 South Seton Avenue, Emmitsburg, Maryland 21727. The report and the photographs, in color, are available on the Administration's Web site at http://www.usfa.dhs.gov

U.S. Fire Administration

Mission Statement

As an entity of the Department of Homeland Security, the mission of the USFA is to reduce life and economic losses due to fire and related emergencies, through leadership, advocacy, coordination, and support. We serve the Nation independently, in coordination with other Federal agencies, and in partnership with fire protection and emergency service communities. With a commitment to excellence, we provide public education, training, technology, and data initiatives.

TABLE OF CONTENTS

OVERVIEW 2

KEY ISSUES 3

REPORT OBJECTIVE 3

FOCUS 4

CASUAL AND CONTRIBUTING FACTORS 4

BACKGROUND 5

WILDLAND FUELS 5

FIRE SERVICE ORGANIZATION 6

STRUCTURAL FIRE PROTECTION 6

FIRE RISK 7

EMERGENCY MANAGEMENT 7

MUTUAL AID 8

WILDLAND FIRE PROTECTION 8

THE 1998 FIRE SEASON 9

DEFENSIVE STRATEGY 10

FIRE SUPPRESSION OPERATIONS 11

MANAGEMENT STRUCTURES 12

TACTICAL COORDINATION 14

COMMUNICATIONS 16

EVACUATIONS 16

RESOURCES 17

TIME TO DELIVER RESOURCES 19

ADDITIONAL ISSUES AND CONCERNS 20

    Fire Behavior 20

    Fatigue 21

    Infrared and Satellite Imagery 22

APPENDIX A: Maps 23

Wildland Fires
Florida - 1998
May - July 1998

Investigated By: J. Gordon Routley

Local Contacts: St. John County
                  Ray Ashton
                  904 823-2527

                  Bunnell FD
                  Chief Gary Hughes (Career Ltd. Daytona Beach)
                  437-7505 Old Hwy 11

                  Volusia County
                  Jim Tauber
                  904-254-4657
                  123 W Indiana Deland 4th floor

                  Ormond Beach
                  Chief Barry Baker
                  904-676-3255
                  Commander Jeff Downs

                  Brevard County
                  Jeffrey Money (Training Chief)
                  407 633-2056

                  Area Commander
                  Larry Wood FDF in Tallahassee
                  850 488-6111

                  Florida Division of Forestry
                  Mike Kuypers
                  Bunnell Division Manager

OVERVIEW

The entire State of Florida was ravaged by an unprecedented series of wildland and urban interface fires during the period from late-May to mid-July, 1998. Almost 500,000 acres were burned, along with more than 150 structures and 86 vehicles, in more than 2,200 individual fires. The total direct and indirect economical impact of these fires will probably exceed one billion dollars. The damage to timber alone was estimated at over $300 million.

The total response to these fires, combining local, State and Federal resources, may be the largest ever assembled in the United States. An estimated total of more than 10,000 fire fighters from across the United States were ultimately involved in the battle to contain the flames. This massive response was required due to the number of fires that were burning simultaneously and the direct threat to dozen of populated communities along the eastern coast of the State. The magnitude and complexity of the operations seriously challenged the capacity of existing incident management systems.

The sequence of events created tremendous challenges for Florida's fire suppression forces, who experienced more than a month of almost continual operations, battling unrelenting flames throughout the State. This was followed by a long weekend of overwhelming fires that threatened to destroy whole communities. Massive public evacuations were implemented as reinforcements from hundreds of miles away converged to join the battle. While thousands of acres were burned and property losses amount to hundreds of millions of dollars, the fire fighters were successful in preventing the urban conflagrations that were feared as the fires reached their peak. No lives were lost and only one serious injury was reported.

In many respects the situation was similar to several wildland interface fires that have occurred in California and other western States. The Florida fires were unusual because they occurred in a State that is not usually considered to be at tight risk for wildland fires, and because of the direct and immediate threat to highly populated areas. These factors once again demonstrate that a change in climatic conditions can create overwhelming fire conditions, in spite of past experience.

The situation is also similar to other major wildland fires in the sense that two different fire fighting components, wildland and structural, had to work together to save lives and property. When a massive fire is moving into a populated area, the only feasible strategy is to identify defensible positions and allocate resources to save the areas than can be safely protected.

KEY ISSUES

Issues	Comments
Unusual and extreme fire risk conditions	The El Niño weather condition created a drought, which elevated the fire risk level throughout Florida to unprecedented levels.
Populated Interface Areas	The rapid development of many communities has created a large number of interface areas. The level of fire risk in many of these areas had not been recognized.
Resource Demands	The magnitude of the situation demanded tremendous resources from throughout the nation. Even with a maximum commitment, it was very difficult to provide sufficient resources quickly enough to meet the demands.
Complexity	The magnitude and complexity of the operations challenged and may have exceeded the capabilities of established incident management systems.
Structural Fire Fighting Resources	There is no national system in place to facilitate a large scale mobilization of structural fire fighters and other resources. The in-State system worked well under extreme circumstances.
Fatigue	The fires fully occupied the local wildland and structural fire services for more than a month, before the most critical period occurred. Many fire fighters had worked to the point of exhaustion.
Aircraft Operations	The magnitude of the air operations was unprecedented and created new challenged in relation to communications, management of air space and operations in populated areas.
Interoperable Communications	The existing communications systems were overwhelmed by the magnitude and complexity of the operations. Interoperability between systems was a major weakness
Planning	Planning must forecast worst case scenarios and anticipate their occurrence. Pre-positioning of resources and establishment of defensive positions well in advance may prove to be invaluable.
Unified Command	Coordination between wildland and structural fire fighting units was often difficult. Unified command structures should include all components that are operating on a fire.
Evacuations	Evacuations of large populated areas require substantial time and resources, in addition to those that are required to fight the fire(s).

REPORT OBJECTIVE

The primary objective of this report is to document experiences that will be helpful to the fire service and emergency managers in planning for situations that may occur in the future. The identification of problems and weaknesses is not intended as criticism. All of the problems that are identified were reported by individuals who were directly involved in the operations and are intended as "lessons learned."

The heroic efforts of the thousands of fire fighters who were involved in these operations have been widely applauded by the citizens of Florida. The citizens who saw the fires are extremely appreciative of the courage and skill that were demonstrated by the fire fighters who were there to protect them.

FOCUS

The focus of this report is primarily directed toward the situation that occurred along the northeastern coast of Florida, between Jacksonville and Orlando, in June and early July 1998. During this time period hundreds of fires were burning throughout the State. This particular area was the most critical in terms of lives and structures at risk, however, the overall situation was much larger and more complex than the aspects that are described in detail in this report.

This analysis also directs particular attention to the operations of structural fire departments and local agencies, which had to operate under severe stress in an unusual and unanticipated set of circumstances. Most of the fire departments that were involved in this situation have very limited resources and had little experience with operations of this magnitude and duration. This experience should provide valuable information for other fire departments that may have to face similar situations in the future.

This report also describes the role of the wildland fire suppression forces, however, these operations are more thoroughly analyzed in other reports. The systems that are used to assemble and coordinate large scale wildland fire fighting operations are sophisticated and well established. While the operation in Florida was one of the largest and most complicated wildland fire fighting campaigns in US history, all of the essential components to conduct this type of operation were already in place. The wildland fire suppression operation was greatly complicated by having to fight multiple simultaneous fires in close proximity to populated and highly vulnerable areas.

CAUSAL AND CONTRIBUTING FACTORS

The most direct contributing factor to this series of events was an unusual weather pattern that caused abnormal patterns of precipitation across most of North America during 1998. The "El Niño" weather conditions, which resulted from elevated water temperatures in a large area of the Pacific Ocean, brought drought conditions to all of Florida, while several normally arid regions of the country experienced above average precipitation.

Florida is not known for major wildland fires. The State has large areas of rich timber lands, which contribute very significantly to the State's economy. While the dense vegetation that is native to most of the State provides an ample supply of fuel, the high humidity and frequent rains keep Florida's wildland fire risk under control in most years. The fires that do occur are usually contained quickly, although the lush vegetation often required major efforts to complete overhaul and final extinguishment.

The events of 1998 illustrate how quickly this balance can change with the forces of nature. The drought index in most of Florida during May, June and early July was over 700 on a scale of 800. Temperatures and humidity readings were similar to those often encountered in the western States. Without their normal moisture content, the abundant pine forests were easily ignited and burned very rapidly.

One previous experience with an unusually dry period demonstrated the vulnerability of these areas to interface fires. On Good Friday, May 17, 1985 a series of fires broke out almost simultaneously and burned into several coastal communities. On that one day, 131 structures were destroyed in the Palm Coast area of Flagler County. The weather conditions at that time were very similar to those encountered in 1998, however, the weather changed within 24 hours and the high fire risk conditions diminished.

The fires of 1998 involved some of the same areas, but were much larger and occurred over a period of almost two months. In the intervening years, the population at risk in those areas has increased tremendously and rapid population growth has created hundreds of miles of new wildland-urban interface throughout Florida. Many of the residents in these areas were unaware of the risk.

The magnitude and duration of the situation that occurred in 1998 greatly exceeded the capabilities of the fire suppression forces that were available. The firefighters were successful in protecting most of the highly populated areas, including some extremely vulnerable communities, from fires that could not be stopped. The fire fighters performed admirably and defended the front lines until a change in the weather ultimately brought relief.

BACKGROUND

Within the past decade Florida has become the nation's 4th most populous State, primarily through the continuing migration of new residents who are attracted by the climate and living conditions. Existing communities are bulging and new communities are developing rapidly in many areas, particularly along the coast lines. Along the east coast this has created long narrow band of developed communities, with wooded rural areas to the west and the Atlantic Ocean to the east.

The fire problem in Florida has become much more difficult to manage as this rapid development has occurred in areas that were previously very rural. Many new communities have been carved out of the State's abundant forests, creating hundreds of square miles of urban-wildland interface zones. It is not unusual for homes in these communities to be surrounded by trees and thick underbrush to within a few feet of the structures. Large developments have been built within heavily wooded areas where the homes and other structures are almost completely surrounded by natural vegetation. Many of these developments have no hydrants or public water systems that can be used for fire suppression.

It is also common in the more rural areas of Florida for individual structures to be built in small clearings, accessible only by single lane roads cut through the woods. Since the fire risk is relatively low in most years, the vulnerability of these structures is often overlooked. When the woods are dried out, the risk level in these areas is extremely high.

WILDLAND FUELS

The natural ground cover in most of these areas is thick pine forest, with a heavy undergrowth of grasses and native plants. If left alone for several years, this material tends to form a thick mass that is almost impossible to penetrate on foot or in wheeled vehicles. These natural fuels are highly combustible, particularly many of the native plants, such as the palmetto, which produce oils and waxes in abundant qualities. Deep-seated fires are common in the undergrowth and are most often controlled by cutting a fire break around the burn area with a plow attached to a tracked vehicle.

This fuel is quite different from the natural cover in many other parts of North America, particularly the northern and western States, which is only susceptible to fires during periods of very low humidity. Because of the waxes and oils within many of the plants, they will burn readily while they are still green and relatively moist. When a drought occurs, the woods become a tinderbox, easily ignited and very difficult to extinguish. The experience and fire growth models that are used in other areas have to be adjusted to successfully predict wildland fire behavior in Florida.

Florida's high humidity, frequent participation and flat terrain usually assist in keeping the fire risk under control. Unfortunately, these characteristics also add to the problem of access into the wildland areas when fires occur. Even in an exceptionally dry year, the ground is often damp or swampy and can only be traversed by tracked or 4-wheel drive vehicles. The thick undergrowth makes access to wooded areas very difficult. Conventional vehicles are restricted to improved roads.

While a 4-wheel drive vehicles can often get close enough to a fire to achieve knock-down with water, tracked equipment is usually required to plow trenches around a fire to achieve full containment. Overhauling a fire in the underbrush usually requires extensive manual labor, as this material may continue to smolder for weeks after a fire has been contained.

The extreme weather conditions in 1998 set the stage for disaster, with dozens of fires that could not be extinguished and more fires being ignited every day. The magnitude of the situation would only be resolved by an end to the drought and the crisis continued until the rain arrived in July.

FIRE SERVICE ORGANIZATION

The forces that battled the Florida fires included both structural and wildland fire fighting organizations. Most of the fires originated in rural and wildland areas, however, the ultimate battle occurred in the streets of built-up communities along the Atlantic coast. Both structural and wildland fire protection agencies were involved in every stage of the campaign.

STRUCTURAL FIRE PROTECTION

The structural fire suppression agencies in Florida include a mixture of county and local municipal fire departments and independent fire districts. These fire departments include some fully volunteer and some fully career organizations as well as combination departments. In most cases the fire departments also provide emergency medical service.

The structural fire departments are also responsible for initial attack on wildland fires within their areas, particularly where the fires are close to populated areas and accessible. The more rural areas are almost entirely protected by volunteers and these departments often respond to more outside fires than structure fires. The rural departments work closely with the Florida Division of Forestry (FDF), which has the primary responsibility for wildland fire protection, and often provide support and assistance to the FDF crews that come in to their areas.

The organization of fire departments varies considerably between different counties. The more rural counties, including Flagler and St. Johns, generally depend on volunteer units, which are organized independently or as components of a county fire service. The independent departments are supported to varying degrees by county funds and services, such as fire prevention and training. Most of the counties provide dispatch and communications for the rural fire departments through a county 9-1-1 center.

As the counties increase in population, the county fire departments often begin to provide career personnel to supplement the volunteers or to have career personnel staffing the stations in more populated areas. Some of the more developed counties, such as Brevard and Volusia, have transitioned to primarily career fire departments in built-up areas, supported by volunteers in more rural areas. There are also a few independent fire districts with career personnel serving built-up unincorporated communities, such as Palm Coast in Flagler County.

Most of the counties also contain incorporated municipalities that operate their own municipal fire departments. Some of the more populated counties, such as Volusia and Brevard, include a series of contiguous municipalities, interspersed with unincorporated areas, all along the coast. In general, the county fire departments are responsible for unincorporated areas, while the municipal fire departments protect incorporated communities, however, some of the municipalities contract for service from a county fire department. The remaining municipalities either operate their own fire departments or contract with a neighboring municipal fire department. The municipal fire departments also range from fully career to fully volunteer.

The municipal departments tend to be more oriented toward structural fire suppression and less oriented toward wildland operations than the county fire departments. The municipal fire departments often operate their own radio and dispatch systems and have limited capabilities to communicate with other fire departments.

Volunteer staffing is a problem in many of the rural areas and particularly problematic in the rapid growth areas. While the populations is growing, it does not provide an adequate supply of residents who are willing to and capable of performing strenuous duty. This is primarily responsible for the trend toward career staffing as areas develop.

FIRE RISK

Until the fires occurred in 1998, wildland and interface fire protection was not considered to be a major concern in most parts of Florida. This can be attributed to the low to moderate fire risk factors in the built environment and the high humidity that limits fire frequency and severity. Major structure fires, as well as wildland fires, are infrequent occurrences. While the residents usually place a high emphasis on emergency medical services, they generally do not view fire suppression as an equal concern.

Limited tax revenues and a desire to keep taxes at a low level also tend to keep fire department resources thinly distributed in most of Florida. In many areas it is not uncommon to find two career personnel staffing a fire suppression company and covering a large geographic area. These limited fire suppression resources are often found in municipal as well as county fire departments.

EMERGENCY MANAGEMENT

Since 1992, when Hurricane Andrew caused unprecedented devastation in southern Florida, there has been a major emphasis on emergency management and on developing mutual aid systems among the fire departments and other emergency response organizations. The primary emphasis in much of this planning has been preparedness for hurricanes and natural disasters, as opposed to major fires, although fire suppression resources are usually an integral part of the disaster response plans.

The Florida Division of Emergency Management coordinates Statewide response to major situations and provides assistance to counties and local jurisdictions, including both funding and resources. The focal point of this system is the State Emergency Operations Center in Tallahassee. Each county has an emergency manager who is responsible for developing plans, coordinating local resources and maintaining links to the State system. The Federal Emergency Management Agency (FEMA) works with the Division of Emergency Management when Federal resources are requested by the State.

MUTUAL AID

The fire service mutual aid system in Florida is operated by the Florida Fire Chiefs Association in cooperation with the State Fire Marshal and is closely associated with the emergency management system. This network is primarily involved in coordinating fire service mutual aid between counties, similar to the California system.

Mutual aid coordinators have been designated and local mutual aid plans have been adopted within each county. These plans provide a system for requesting and coordinating assistance and managing incidents among those fire departments. The mutual aid coordinator is usually a chief officer from one of the fire departments in the county.

When a situation exceeds the capabilities of a county's own resources, the county coordinator can request assistance through the Emergency Operations Center in Tallahassee. The EOC then contacts the mutual aid and coordinators in other counties that have resources available and authorizes them to dispatch units to fill the requests.

The mutual aid resources are usually dispatched as task forces or strike teams, assembled from the fire departments within the county. The local fire departments are reimbursed by the State for this assistance.

WILDLAND FIRE PROTECTION

The primary responsibility for wildland fire protection rests with the Florida Division of Forestry (DOF) Florida's forests are valued as a natural resource, both to protect the natural environment and for the revenue produced by timber as a commercial product of the sate. This agency has resources distributed throughout the State, assigned to regional management units, to manage the forests and provide fire protection for thousands of square miles of public lands.

The local structural fire departments and DOF units generally have very close established relationships at the local and regional level. The local fire departments usually provide an initial attack on wildland fires in the areas they can reach, while the DOF resources assume the responsibility for fires that require additional resources or prolonged efforts to control. Once the DOF units arrive and assume responsibility for a fire, the local forces usually take a supporting role for as long as they are needed.

The DOF resources include hand crews, wildland engines, tractors with plow attachments and bull-dozers, as well as a few helicopters. The tractor/plows are routinely used to isolate fires that are burrowing through the underbrush. The overhaul operations at individual fires may take several days, however, once the fire is contained, the hand crews can usually limit their operations to daytime hours.

The Florida Division of Forestry is involved in the Nationwide Wildland Fire Coordination (NWFC) Group and works closely with other State and Federal agencies that provide fire protection for public lands. This system is coordinated through the National Interagency Fire Center (NIFC) in Boise, Idaho. A separate cooperative agreement, known as the Southeastern Compact, exists among several bordering States to provide mutual assistance for a variety of emergency situations, including wildland fires.

The US Forest Service is responsible for wildland fire protection in the areas of Florida that are under its jurisdiction, including the Ocala, Apalachicola and Osceola National Forests. The Fish and Wildlife

Service is responsible for protecting significant coastal areas within the State. These and other Federal agencies work closely with the Florida Division of Forestry.

In addition to creating new interface areas, the increasing urbanization of Florida has made wildland fire risk mitigation more challenging. Prescriptive burning, which is an important program to thin the underbrush and interrupt fuel continuity, has been severely curtailed because of smoke complaints from developed areas.

Until 1998, the general population was not greatly concerned with the risk or wildland fires. While fires do occur in the wooded areas, the humidity usually keep them from becoming large enough to pose a significant threat to built-up areas. Regulations requiring brush clearance around structures have not been adopted or enforced in many areas, because interface fires have not been identified as a major risk factor by local jurisdictions.

THE 1998 FIRE SEASON

The series of fires that ravaged Florida in 1998 began in mid-May and continued through the early part of July. During this period approximately 500,000 acres were burned in approximately 2,200 separate fires. The causal factors that were responsible for these fires began to develop during the previous year.

During the fall and winter of 1997-98 the El Niño weather pattern caused major worldwide deviations from normal temperatures and precipitation. The impact on Florida was an unusually wet winter, followed by a very dry spring, and an early summer with higher than normal temperatures.

The winter rains promoted unusual growth of the low vegetation and also curtailed the established program of prescriptive burning, which is intended to thin this fuel. The fire risk conditions then increased steadily as the woods dried out and water tables dropped, making the entire State a proverbial tinder box. These conditions developed in April and prevailed through May and June, with record high temperatures, low humidity and no precipitation.

By late May, fire departments and wildland fire forces throughout the State were busy responding to numerous brush and wood fires that were breaking out day after day. A large proportion of these ignitions were caused by dry lightning, while other were man-caused, including several suspected cases of arson. This time of the year is known for frequent thunderstorms, however, in 1998 the storm cloud brought the lightning without precipitation. Some of the thunderstorms were triggered by the thermal plumes from large fires.

The fire problem was particularly severe in the northern counties and the Florida panhandle, although there were outbreaks in almost every part of the State. The largest number of fires occurred between along the northeastern side of the State, between Orlando and Jacksonville. This area includes Brevard, Osceola, Orange, Seminole, Flagler, Volusia, St. Johns and Duval Counties.

The fire activity continued to increase during the first week of June with increasing temperatures and no relief from the drought. Fires were breaking out in most of the northern Florida counties, particularly in St. Johns and Duval Counties, where several fires threatened the suburban areas of St. Augustine and Jacksonville. Major fires in the Apalachicola and Osceola National Forests caused a major commitment of US Forest Service resources.

During this period the Florida Fire Chiefs' Statewide mutual aid system was processing frequent requests, moving fire department resources from county to county on a daily basis. Additional wildland resources were deployed from neighboring States under the Southern Forest Fire Compact to assist the DOF crews in rural areas.

The first crisis situation occurred between June 6th and 8th, as several new fires broke out between Jacksonville and Orlando. On June 6th, 30 structures were lost in Geneva in Seminole County. On the same day another 20 homes were lost in the Seminole Woods area of Palm Coast, in Flagler County, while all of the surrounding counties were contending with multiple outbreaks. On June 7th Governor Lawton Chiles declared a "State of Emergency" and mobilized the Florida National Guard to assist in the wildland fire fighting effort.

The number of fires continued to increase for the next two weeks. In many cases it was impossible to fully contain and overhaul the fires with the extreme weather conditions and the number of new fires that were breaking out in rapid succession. The normal tactic of knocking down the flames, then isolating the burning undergrowth by plowing around the fire, was ineffective with the drought conditions. Also, as more fires were reported, crews often had to leave fires that were contained, but not fully extinguished, and move on to attack new fires. Full control and extinguishment would only be accomplished with a change in the weather.

It was evident that the fire risk would continue to increase until a change in the weather brought higher humidity and enough rain to end the drought. This increased the level of concern for the overall situation and prompted many of the agencies to anticipate even greater problems in the following weeks. The State and county governments began to implement their emergency operations plans and experienced overhead teams from the national wildland fire fighting network were brought in to manage the operations in several areas. Fire fighting aircrafts were redeployed from several western States to join the battle in Florida.

DEFENSIVE STRATEGY

As the number of active fires continued to increase over the next two weeks, the overall strategy in several counties began to shift from offensive to defensive. The available fire suppression forces were fully committed to attacking new fires that posed a direct threat to populated areas, while fires in isolated areas were given a lower priority. Planners began to identify built-up areas that would be vulnerable several days or weeks in the future, if the fires could not be stopped. The planning concentrated on establishing defensive lines to keep the fires out of developed areas.

A Presidential disaster declaration was signed on June 18th to authorize direct Federal assistance to fight the fires. Additional overhead teams and wildland fire suppression resources from several more States were sent to Florida to assist the DOF and US Forest Service crews. Aircraft was also redeployed from bases in western States to bases in Florida.

Between June 19th and 22nd, dry thunderstorms ignited more than 80 additional fires per day in the same general area. Most of the fires in accessible areas were quickly controlled by ground forces and aerial attack, however, several other fires in remote areas continued to burn. The drought conditions continued through the rest of June and the battle went on, day after day, burning thousands of acres and wearing out the personnel and equipment that were available.

On June 21st a unified Area Command was established at the State EOC to direct the Statewide response to the growing fire situation. This command included the Florida Division of Forestry, Division of Emergency Management, US Forest Service and FEMA. In the following days eight "complex commands" were established to manage operations in different parts of the State, reporting to the Area Command. A Type 1 or Type 2 incident management team from the nationwide wildland system was assigned to each of these complexes.

The situation reached its most critical stage during the last days of June and the first week of July, when several of the fires in rural areas began to converge, forming much larger fires. A series of very large fires developed in the area west of Interstate 95, in Brevard, Volusia and Flagler counties. A sudden flare-up occurred on July 1st, when hot dry winds pushed the flames toward the populated communities that line the coast. This situation prompted an unprecedented response of both wildland and structural fire fighters and supporting resources into Florida from across the United States.

On July 1st and 2nd the fires jumped across I-95 and other natural firebreaks in several locations as they moved toward the east. The fires threatened to sweep into several urban areas and residential subdivisions. An intense battle raged for three days, as the flames consumed dozens of structures. In Volusia County the fires reached built-up areas of Daytona Beach and Ormond Beach. On the afternoon and evening of July 2nd flames consumed 51 dwellings in Palm Coast in Flagler County. In northern Brevard County 36 homes were destroyed during the same period.

Large scale evacuations were ordered in several communities. More than 15,000 residents were evacuated from areas of Ormond Beach that were in the path of the fires. On July 3rd the entire civilian population of Flagler County, more than 45,000 residents, were ordered to evacuate in the face of a potential fire storm.

The threat began to subside on July 4th as the weather conditions finally changed. Increasing humidity and relief from the dry winds allowed fire fighters to hold the lines and end the crises. The first rain fell on July 5th and more rain followed on the succeeding days. Over the next two weeks all of the remaining fires were contained, overhauled and fully extinguished.

FIRE SUPPRESSION OPERATIONS

This series of fires that occurred in Florida in 1998 overwhelmed the capabilities of State and local resources in both scale and duration. The fire control efforts, which continued for more than two months, from mid-May to mid-July, combined both wildland and structural fire fighting resources. Some of the fires that originated in remote areas eventually burned into built-up communities, jumping interstate highways and natural features that would normally be expected to contain a wildland fire. The battle eventually became a street fight, with fire fighters trying to protect individual structures in the coastal communities from an advancing wall of fire.

Beginning in May, both wildland agencies and structural fire departments faced a rapid increase in the frequency and severity of fires. Local fire departments, particularly in the rural areas, were occupied day after day, responding to new outbreaks and often assisting the Division of Forestry crews with fires that had already been burning for several days. This placed a tremendous strain on many of the rural fire departments that normally operate with minimal resources and rely heavily on volunteers. Some of the volunteers were released from their regular jobs and became "full-time" volunteers, while others had to limit their participation to maintain their employment. In the career and combination departments, off-duty personnel were called back on overtime to duty, to staff extra units.

Initially, the local mutual aid plans were activated within individual counties, often mobilizing municipal fire departments to back-up the rural fire departments. Water tenders and brush units were staffed and spent long days fighting fires, concentrating on keeping the flames away from structures and other valuable property.

As the number of fires increased, requests for mutual aid from other counties were directed to the Florida Fire Chiefs network. The EOC in Tallahassee assigned these missions to the mutual aid coordinators in counties that had resources available. Individual resources from fire departments within these counties were assembled into task forces and strike teams, which were dispatched to the counties that had requested assistance.

Assignments of strike teams and task forces from one county to another, often involving travel of 50 to 100 miles, became daily occurrences. In many cases the reinforcements were requested in the mornings and released in the evenings, when fires subsided, and new requests were initiated on the following days. As the situation progressed, the mutual aid resources from distant counties were often held overnight and operated for two or more consecutive days.

The mutual aid system was strained by the circumstances that several counties in a large region were simultaneously experiencing multiple fires and requesting mutual aid. During June, all of the 67 counties in Florida were at some level of emergency status due to the fire situation. In many cases a county that was able to send units out in the morning had to request mutual aid for fires in their own county later in the day. Units that were exhausted after spending hours on a fire in a neighboring county sometimes had to respond back to attack a new fire that had broken out while they were gone. As this went on, day after day, several counties determined that they would have to decline mutual aid requests and retain all of their units for local protection.

As the requests for mutual aid increased, units from more distant counties were assigned to fill the requests. It often took two to three hours for the first units to reach a county that had requested assistance in the morning. It could take even longer for reinforcements to respond to requests later in the day, as all of the resources in nearby counties were already committed.

The increasing number of fires also quickly exceeded the capabilities of the Division of Forestry. As the number of fires increased, Florida began to request assistance from other States, as well as Federal resources. Assistance was initially requested from the neighboring southeastern States. As the situation continued to escalate, more Federal and State resources from across the county were requested from the National Interagency Fire Center in Boise, ID. This resulted in a gradual build-up of wildland forces through most of June.

An unprecedented number of aircraft were assigned to fight the fires in Florida. During the peak period all of the available fixed wing attack aircraft and large helicopters in the lower 48 States were committed to this operation. A total of 156 aircraft were assigned to six bases.

MANAGEMENT STRUCTURES

During late May and into June, most of the fires occurred in rural areas where the jurisdiction was shared between county fire departments and the Florida Division of Forestry. Following established procedures, the local fire departments would make the initial attack on fires they could reach, then turn the fires over to DOF units when they arrived. In the great majority of cases the local fire fighters were seldom involved with a fire beyond the first day.

During these early stages, strategic decisions related to individual fires were generally made at the local level, often through informal relationships between local fire officials and DOF Division

Managers who had worked together for several years. The DOF geographic divisions usually encompass multiple counties and local jurisdictions, so each manager had these relationships with several different fire departments.

As more and more fires were ignited, the Division Managers had to coordinate operations on several fires and maintain multiple liaisons with several local fire departments at the same time. The area commanders were very mobile, attempting to monitor conditions, allocate resources and direct operations on several fires at the same time. As the DOF resources were stretched, the local fire departments had to take on an increasing role in attacking fires and supplementing the wildland forces, often dealing with multiple fires on successive days.

After a state of emergency was declared by the Governor, in early June, National Guard resources became available to support the DOF operations, however, the number of fires was still increasing. When the Emergency Operations Center in Tallahassee was activated to manage the overall situation, allocating resources based on the risk factors for each fire, it was initially operated by a unified team involving Florida DOF and US Fire Service personnel, as well as the Georgia Forestry Commission. The coordination and allocation of wildland resources was moved up to this level, while the FDF regional manager continued to direct operations on the individual fires in their assigned areas and coordinate activities with the local fire departments.

As the threat increased, several counties also activated their own emergency operation centers and implemented more structured incident management organizations for their own forces. In several counties, the county fire departments established formal unified commands with the local DOF managers - this varied significantly from county to county, depending on the local organizations and resources.

As the situation continued to grow, overhead teams with experienced command and staff personnel were dispatched to Florida. The first overhead teams came from the southeastern States. These teams were initially assigned to assist with overall coordination and resource allocation at the State EOC and to some of the major fires.

The overall organization changed radically in late June and early July, after the Presidential Disaster Declaration was signed. The Area Command was established in Tallahassee and additional Type 1 and Type 2 overhead teams from the nationwide system were deployed. Tremendous operations and support resources began to move into Florida from across the country.

The overhead teams established more sophisticated incident command structures, setting up the eight "complex" organizations to manage operations within designed regions. The incident management team (MT) assigned to each of these complexes assumed responsibility for all of the fires that were burning within a large geographic area. The previously assigned DOF commanders and staff personnel were absorbed into a unified command structure in each area and a large scale base of operations with hundreds of personnel was established in each region.

The new command structure brought the State and Federal wildland fire fighting forces together, with shared command responsibilities over wildland operations, however, most of the local fire departments were not included in this structure. While each Incident Management Team was responsible for all of the wildland fires that were burning within a large geographic area, the local fire departments were still responsible for initial attack on new fires.

The structural fire departments were also directly concerned with defending their own geographic areas. The county and local fire departments worked within their county mutual aid systems and the Florida Fire Chiefs' Mutual Aid network to obtain resources to fulfill this mission. They also worked closely with the local emergency managers to ensure that their areas were prepared for any potential encroachment of the fires into populated areas. The local agencies could request resources and logistical support from the regional wildland commands, as well as from the State EOC.

The area assigned to each "complex" command component encompassed all or parts of multiple counties. As a result, several different fire departments were involved in liaison relationships with none more than one command. Response areas were usually split consistent with geographic features, such as a river or highway, that did not necessarily coincide with county or fire department jurisdictional lines. Some of the fires burned across county and municipal boundary lines into different local jurisdictions.

This created a confusing situation, as some of the local fire departments were responsible for structural fire protection in areas that were split between two different wildland commands, with their command posts located 50 miles apart in different counties. The incident command teams that were assigned to the complexes were usual from distant States and Federal agencies and were unfamiliar with the local individuals and fire service organizations.

With numerous fires and few command officers, it was very difficult for the county and municipal fire departments to assign full time liaisons to the wildland command posts, while continuing to manage their own operations and resources. In addition, the local fire departments had a key role to play in each of the county and municipal emergency operation centers that were operating.

The command staffs of these departments were extremely busy during this period.

TACTICAL COORDINATION

The level of strategic and tactical coordination between wildland and structural forces varied significantly from one county to another. Although they were often working side by side on the same fires and supporting each other's operations, the two components were usually managed and directed through parallel systems.

During the early stages, the tactical coordination between units in the field was informal and often based on established relationships between individuals. The DOF managers and local fire officers made strategic and tactical decisions based on the situations they encountered and the resources that were available at that time and place.

Most of the fires originated in rural or sparsely populated areas, where wildland tactics were appropriate. Most of the local fire fighters in these areas were trained and equipped for initial attack on small brush fires and were able to provide support to the wildland forces, but they were not prepared to conduct large scale wildland fire fighting operations. Most of their efforts were directed toward attacking fires that were close to roads and structures and protecting built-up areas from encroaching flames.

As the situation progressed, more resources were assigned from structural fire departments in built-up areas to assist on the rural fires. These fire fighters were generally less prepared to fight wildland fires, often lacking the necessary training, experience and equipment. The wildland fire fighters were equally unprepared for structural fire fighting.

Looking ahead, the DOF managers and local fire department command officers began to identify areas that would be vulnerable days or weeks in the future, if the weather did not change. They began to plan defensive tactics that could be employed to protect vulnerable built-up areas. In some of the counties the fire departments coordinated local efforts to cut fire breaks and clear brush around these areas.

The coordination between wildland and structural forces became more complicated as the fires became larger, more resources were deployed and the management organizations became more sophisticated. As the wildland incident commands became more structured, with overhead teams assuming command of the major fires and complexes, their planning became more strategic. In many cases the local fire departments were unaware of the strategic plans that were being adopted by the wildland forces. Communications between the wildland and structural forces were often very limited, particularly at the tactical level--units working on the same fire often had no ability to contact each other by radio.

As the situation became critical, the commanders were faced with growing and potentially over-whelming fires that were threatening to overrun built-up areas. At this point the wildland forces placed a priority on establishing strategic defensive containment lines, using geographic features such as major highways, and assigned their resources to make their stands at those locations. The local fire fighters were often committed to defending their own jurisdictional areas, protecting individual structures and neighborhoods. As a result, the structural fire fighters sometimes found themselves out in from of the lines, attempting to defend individual structures, while the assembled wildland resources stood by, waiting for the fire to come to them. Both strategies may have been appropriate for the circumstances, however, there was much consternation over the lack of coordination.

When the fires flared up on July 1st and began to move toward the coast, several highly populated areas were directly exposed, including several communities that are protected by municipal fire departments. This placed several more fire departments on the front line preparing to protect their own communities. In some cases the county and municipal fire departments established unified command structures among themselves as they prepared to defend their territory. The preparations included cutting fire breaks around subdivisions and pre-positioning strike teams to defend vulnerable neighborhoods. The wildland forces were involved in many of these actions, but remained under a separate command structure.

During the most critical period, the battle was fought in the streets of built-up communities, including Palm Coast and Ormond Beach, trying to keep the flames away from individual structures. This action involved both wildland and structural fire fighters, often with a mixture of tactics and equipment, including aircraft. During this period the wildland and structural forces tried to coordinate their strategy and tactics by utilizing the same geographic references and organizational designation systems, however, this often proved to be very difficult.[1]

---

[1] The problems encountered in Brevard County illustrate the types of situations that occurred in several areas. Most of the municipal fire departments in Brevard County utilized the County's 800 MHz trunked system, which was designed with three independent transmitter sites, located in the south, central and northern part of the county. Each department's radios were programmed to operate primarily from one of the three sites and only a limited number of talk groups could be used with the other tower sites. All of the radios had to be reprogrammed to become fully operational on all three sites and several additional radios were acquired to deal with the crisis.

COMMUNICATIONS

The magnitude of the situation and the number of different agencies that were involved created tremendous communication challenges. The level of activity over a large area, the number of different fires and the number of units operating simultaneously overwhelmed most of the radio systems that were in place before the fires. Some of the agencies supplemented their radio systems with cellular telephones and made emergency acquisitions of additional portable radios, but these options were often limited by infrastructure capacity and the operational capabilities of each radio system.

The designs of these systems never contemplated operations of this magnitude and they would not accommodate the demand for communications. The pall of smoke produced by the fires also interrupted microwave links between communication centers and remote transmitter and receiver sites, temporarily disabling major system components, sometimes for several hours at a time.

Interagency communications were even more difficult, as units from different jurisdictions with incompatible radio systems were frequently assigned to the same fires. The few channels normally reserved for mutual aid and interagency communications were quickly overwhelmed. Mutual aid units were often unable to communicate with the local communication centers or with other units operating around them. Numerous instances were reported where units could not be contacted for several hours.

Mutual aid units coming into the area had to be directed to a check-in point where they could be staged or directed to an assignment. Without reliable communications, they could not be redirected, call back for information or request assistance unless they found a unit that did have a functional communications link. Several units reported that they were commandeered or found critical situations and went to work before they could find their check-in location and only reported-in officially several hours late. Some units also found themselves in areas that had been evacuated, because they had no means to maintain an awareness of the overall strategy.

Incompatibility of radio systems between wildland and structural fire fighting units was a particular problem. In many cases the supervisors and field command officers had to find each other and meet face-to-face to coordinate their activities. Safety is seriously compromised when units that are working on the same fire are unable to communicate with each other. This coordination is also extremely important when aircraft are operating in the proximity of units on the ground.

Radio caches brought in by the Federal teams were put to good use, although they were limited by the flat terrain until tower sites could be acquired and base station antennae could be erected. With the number of simultaneous operations in close proximity to each other, even these systems proved to be inadequate.

Several mobile command post units were utilized to coordinate operations within limited geographic areas. In many cases the mobile command posts could communicate with units in a limited geographic area and maintain radio or telephone contact with higher level command centers.

EVACUATIONS

During the most critical period, during the first week of July, the flames were threatening to sweep into dozens of populated coastal communities along a front more than 100 miles wide, from Brevard County to Flagler County. The total population of the areas in the potential path of these fires was more than 500,000 including the Daytona Beach metropolitan area in Volusia County, a series of adjoining communities that stretches approximately 20 miles from Ormond Beach to Port Orange.

In addition to the year-round residents of this area, an estimated 200,000 visitors were expected in the area over the 4th of July weekend to attend a NASCAR race at Daytona International Speedway. The race track is on the western edge of the city, which was directly in the path of the approaching fires. On July 1st and 2nd the flames were threatening structures within a mile of the Speedway and all nearby residents were evacuated. A decision was made on July 2nd to cancel the race and to advise all of the expected visitors to stay away from the Daytona Beach area. The race was rescheduled for a date in October.

When the flare-up occurred on July 1st, the fire departments in Ormond Beach, Daytona Beach and several other communities identified the areas that were in danger and worked with their emergency management agencies to order mandatory evacuations. Approximately 15,000 residents were evacuated ahead of the advancing flames. Many of these residents were accommodated in temporary shelters, which were located in safe areas, while most left the area in their own vehicles and found refuge elsewhere.

To the north of Daytona Beach, another group of major fires was burning in Flagler County. Approximately 50 structures were lost in the large unincorporated community of Palm Coast on the afternoon and evening of July 2nd when the wind pushed the flames and burning embers into populated areas. The following morning reconnaissance reports indicated that these fires had the potential to join together and create a giant firestorm. If this had occurred, it could have wiped out Palm Coast, the town of Bunnell and several similar communities. A decision was made to evacuate the entire population of Flagler County, more than 45,000 residents, in the face of this threat.

The major roads through Flagler County, Interstate 95, US1 and Florida A1A, all run north and south, parallel to the coast and within a few miles of the Atlantic Ocean. The evacuation required all of the residents to use these roads to travel either north, into St. Johns County, or south into Volusia and Brevard counties, all of which were also at a high risk. The evacuees were directed through these counties to either Jacksonville to the north or the Orlando area to the southwest to get them out of the endangered area. These residents were unable to return to their homes for four days.

The smoke and flames caused a 125 mile stretch of Interstate 95, the major east coast transportation artery, and several other major highway to be shut down for several days. Several different fires crossed I-95 in different locations as the fires moved to the east. This caused a huge backed-up of traffic north and south of the area.

RESOURCES

The events that occurred in Florida during June and July resulted in the nation's largest deployment of wildland and structural fire fighting resources. More than 10,000 fire fighters were involved in the operations, which utilized almost all of the deployable wildland fire fighting resources in the United States. The air operation was the largest ever conducted. It is also believed to be the largest commitment of structural fire fighters to a wildland interface situation.

As the number of fires increased, the demand for resources also increased. With wildland fires burning in most Florida counties, the requests for both wildland and structural resources often exceeded the immediately available supply. On many occasions the requests had to be prioritized at the State level, as all of the requests could not be filled. This was a critical problem when the major flare-up occurred on July 1st, as the demand for resources during the next 72 hours greatly exceeded the supply.

Two parallel systems were used to obtain and deploy wildland and structural fire fighting resources. During the early stages, wildland resources were requested by the DOF managers through their own organizations. As the situation escalated, the overall wildland command and resource allocation shifted to a multi-agency command that was established at the State EOC in Tallahassee. As the overhead teams established regional complex commands, the resource requests were initiated at this level and directed to the Area Command, which was also located at the State EOC.

Throughout June and into July, the established systems were used to mobilize additional wildland resources from across the United States as well as some foreign resources. All requests for out-of-state assistance were coordinated through the National Interagency Fire Center in Boise, Idaho, using the established national system for wildland fire fighting resources. Florida National Guard resources were also utilized to support and supplement the wildland forces.

Most of the structural fire department resources were requested by the county mutual aid coordinators through the Florida Fire Chiefs network. During most of June, these requests were filled by in-state resources, with some fire department units traveling long distances within Florida. Fire departments throughout the State were committed to this activity.

When the fires began to move toward the coast, several municipal fire departments, which had been providing mutual aid to the unincorporated rural areas for several weeks, had to prepare to defend their own interface zones. This created a sudden increase in the demand for structural fire fighting resources, as well as wildland resources, from the counties and municipalities that were in the path of the fires. Also, by this time hundreds of fire fighters were at the point of exhaustion after weeks of strenuous duty. These factors prompted the huge mobilization of resources that occurred during the first week of July. The requests included all types of fire fighting and logistical resources, including both wildland and structural fire fighters and all types of apparatus and equipment.

On July 1st and 2nd, urgent requests for structural fire fighting resources went out from the State EOC, through the Division of Emergency Management, to Federal agencies and to other States. Most of these requests went to the emergency management agencies in different States, which passed the requests on to local fire departments and fire service organizations. Within 24 hours fire department resources were being assembled in more than a dozen States and transportation was being arranged for hundreds of fire fighters who had volunteered to join the battle.

In addition to the official requests for assistance from the State EOC, several specific requests for assistance went out from individual fire departments in the endangered area to fire departments in other States. Many of these requests were based on previously established relationships, including informal personal relationships among individuals.[2] Several fire departments quickly assembled contingents and responded to these requests. A significant number of individual fire fighters from other States also made their own way to Florida to volunteer on the spot.

As all of these resources arrived in Florida, the personnel and equipment had to be divided-up between the wildland forces and the structural forces, depending on their training and assigned to different areas. The assistance came in many forms from different sources. Some units arrived ready for action, while others arrived with minimal equipment or days ahead of their vehicles. All of the personnel had to be integrated into the system according to skills and qualifications and assigned where they could be used effectively.

Most of the wildland fire fighters arrived in organized crews, which were easily deployed to the various commands and integrated into the wildland organization. The largest contingent of wildland fire fighters were assigned to the Bunnel Complex in Flagler County and the Orlando Complex in Brevard County. These two complexes managed wildland operations in the seven counties that were experiencing the most critical situation from their base camps in Bunnell and Cocoa.

Hundreds of structural fire fighters, both career and volunteer, most of whom were not trained or equipped for wildland fire fighting, also arrived in the Jacksonville and Daytona Beach areas. The structural fire fighters had to be organized and deployed to either assist or relieve other structural fire fighters. Management systems had to be adapted quickly to log all of these fire fighters and resources into the operation and to deploy them where they were most needed. Some of these groups arrived in time to fight on the front lines as the fires encroached on populated communities. Many more arrived on the following days, as the crisis was waning, in time to provide needed relief for the exhausted fire fighters who had been on the front line for days or weeks.

The arrival of so many more structural fire fighters also created huge logistical challenges. Accommodations and support services had to be provided for hundreds of newly arrived fire fighters, in addition to the hundreds who were already there. A huge camp was established in the infield at the Daytona International Speedway as a base of operations for the structural fire fighters.[2]

The decision to cancel the NASCAR race had a fortuitous side effect, as the hotels and restaurants in the Daytona Beach area suddenly had large stocks of excessive food and other supplies. Most of these provisions were donated to the fire fighting effort by a grateful community and many of the fire fighters were accommodated in beach front hotel, free of charge.

TIME TO DELIVER RESOURCES

One of the major problems experienced in the Florida fires was the time it took for needed resources to be delivered. Under routine circumstances, structural fire departments usually think in terms of minutes for assistance to arrive. The wildland fire fighters in Florida generally anticipate that it will take hours for their resources to arrive. These times became considerably longer as the increasing number of fire demanded more and more resources.

During the first part of June, the delay in delivering structural resources was usually measured in hours, as task forces and strike teams had to be assembled in counties that had resources available and dispatched by highway to the fires. The lag time for obtaining wildland resources was generally measured in days, as all of Florida's resources were committed and assistance had to come from other States. As the demands increased, both of these times lengthened, as structural units had to respond longer distances within Florida and the wildland resources were dispatched from more distant States.

Wildland fire fighting agencies across the nation are networked and fully prepared to respond to distant States on short notice, however, there is no equivalent system in place for structural fire fighting resources. When the critical situation developed around Daytona Beach and in Flagler County on July 1st, the EOC in Tallahassee began to contact emergency management agencies in other States to determine if structural fire fighters were available. Several States responded back to Florida within hours that they could provide personnel from fire departments in their States.

---

[2] As an example, Ormond Beach requested and received assistance directly from Fairfax City, Virginia. This was a direct result of Ormond Beach's new fire chief having previously been a member of the Fairfax City Fire Department.

The official requests from the Governor's Office to these States, under the Emergency Management Assistance Compact, were transmitted on July 2nd. At that time the State emergency management agencies began to authorize the local fire departments to prepare to respond to Florida, while the State agencies worked with FEMA to arrange transportation for the personnel and their equipment. All of this was occurring over the July 4th holiday weekend.

It took 24 to 28 hours for most of these units to be assembled and for air transportation to be arranged. The first out-of-state fire fighters began to arrive on July 4th and were quickly deployed. More help arrived on July 5th and subsequent days, after the critical period had passed. The additional fire fighters were still very valuable, particularly to relieve exhausted crews, however, most of them did not arrive in time for the major battle.

Many of the structural fire fighters were not trained in wildland fire fighting and those who were qualified were unfamiliar with the characteristics of Florida's natural fuels. Also, most of them came with protective clothing designed for interior structural fire fighting, which was not suitable for outdoor operations in the 100 degree temperatures they encountered in Florida. In some cases it took several days to provide an orientation and to procure and issue wildland protective equipment to these fire fighters, before they could be utilized effectively. The logistics functions also had to fee and house these fire fighters and provide ground transportation.

ADDITIONAL ISSUES AND CONCERNS

Fire Behavior

Throughout this series of fires, the fire behavior was very difficult to predict, particularly for wildland fire fighters who were unfamiliar with the fuel conditions and shifting coastal winds encountered in Florida. The mathematical models that are used to predict fire behavior had to be adjusted to account for the very different characteristics of the fuel and its ability to burn in a relatively moist environment.

Fires in thick underbrush fuel are usually very difficult to overhaul, however, the wood fires that normally occur in Florida can usually be fully contained by knocking down the flames with water and then cutting a fire break around them with a plow attached to a tractor. The underbrush would often continue to smolder for several days, however, the fire would be contained within the perimeter. Due to the low humidity and fuel moisture content in 1998, the tangled underbrush material would often continue to burn and eventually reignite the trees and brush, allowing the fire to jump over the fire break. It was almost impossible to fully extinguish these fires without a soaking rain.

Wildland fires usually tend to diminish at night and begin to regain their energy with the morning sun, reaching a peak during the afternoon. In Florida, under normal conditions, most fires are reduced to smoldering at night and can often be overhauled in the morning. As the temperatures increased and fuel moisture content decreased during June, the fires became energized earlier in the day and burned later into the evenings. In early July, as the situation reached its peak, some of the fires continued to blaze through the night.

In the coastal areas the winds shifted frequently and often pushed the fires in different directions during different period of the day. A dry wind would push a fire rapidly in one direction for part of the day, then a moist offshore breeze would slow its pace and reverse its direction. Some of the fires burned back and forth for days, slowly expanding within the same general area. Several fires joined together, while others branched out in different directions to form multiple heads.

Major flare-ups occurred during periods of extremely hot and dry weather, particularly with a strong wind from the west. During these periods huge flame fronts were created and ignitions from embers were reported more than one half mile downwind. Under these conditions the fires easily jumped across natural fire breaks, including Interstate 95, and could have crossed the Intracoastal Waterway. The almost flat terrain had very little impact on fire spread, but also provided very few natural fire breaks.

The situation was recognized as being beyond the capability of fire suppression forces, as long as the dry weather continued. It was a matter of time, whether the rain would come before the fires could reach and overwhelm populated areas. Each day the fire fighters hoped for rain that did not come and the situation became more critical.

Fatigue

The fire fighters of Florida fought a war of attrition that went on for almost two months. Fatigue became a major factor, as units were often committed to fighting fires for 14 to 16 hours and sleep before returning for another full day of operations. In some cases they slept in their vehicles and resumed operations at dawn.

During campaign operations, wildland fire crews are usually committed to one fire at a time and normally operate on rotating shifts. A crew might operate for 12 hours, then shut down for the night while a second shift relieved them. This type of work cycle was virtually impossible to accomplish for the DOF crews and local fire fighters for more than a month.

The local fire departments often had to contend with multiple fires and also had to maintain their 24 hour capability to respond to any other emergency incidents that occurred in their jurisdictions. In many cases, after a full day of fighting fires, they were called out again during the night for new ignitions or rekindles of old fires, as well as for medical and rescue incidents. Many of the volunteers temporarily gave-up their regular employment and became full time fire fighters until the situation was controlled.

It was not unusual for units to travel 40 or 50 miles to work on one fire, then to be directed to turn around and respond back to the area where they had started for another fire. When they could shut down operations for the day, they still had to return to their stations to service their apparatus and equipment. On busy days some of the off-road vehicles logged 150 to 200 miles on the highway.

Command officers and staff support personnel were often on virtually continuous duty, napping for an hour or two when there was opportunity. At command posts and emergency operation centers the planning and logistics functions were active all day and most of the night. The daily decommitment of resources often overlapped with planning and preparation for the next day's operations.

As the events reached their most critical stage, most of the fire fighters were already exhausted, supplies were running low and their apparatus and equipment needed maintenance in spite of the fatigue factor, many units continued to operate for several days, working all day and all night, trying to hold the fires at the edge of built-up areas or conducting a house to house battle. The most important function for many of the out-of-state personnel, when they arrived, was to relieve the exhausted crews that had been on the front line for weeks. In several cases crews from far away jurisdictions were assigned to staff fire stations, while all of the regularly assigned personnel got their first good rest in more than a month.

Infrared and Satellite Imagery

During the weeks that Florida was experiencing this devastating series of fires, the use of infrared (IR) imagery proved to be very valuable. While a blanket of smoke covered hundreds of square miles, IR images allowed planners to map the locations, extent and progress of fires to see if they were growing and to determine how quickly they were moving and in which directions. The IR scans also provided a very quick and accurate capability to detect new ignitions in remote areas.

The initial use of IR technology was from aircraft flying over the area to survey the fires. Engineers at Kennedy Space Center at Cape Canaveral, which is in close proximity to the fires, observed that the fires were visible on satellite IR images and made this technology available to support the fire fighting operations. The satellite images allowed for the entire area to be scanned on a single pass and for the resulting images to be immediately available via the Internet. The satellite IR images provide very high resolution capability and are fully compatible with digital mapping systems and Global Positioning System (GPS) technology.

The increasing use of GPS to map wildland fires has become very valuable for planning and directing operations and greatly increased the efficiency of aircraft operations. Aircraft can be guided to very precise locations by land based units equipped with handheld GPS units. The same technology can be used to direct ground units to the locations of fires detected by aerial observation or IR scans from aircraft or satellites. The maps can also be transmitted electronically to different locations, including field units with mobile data terminals.

Maps

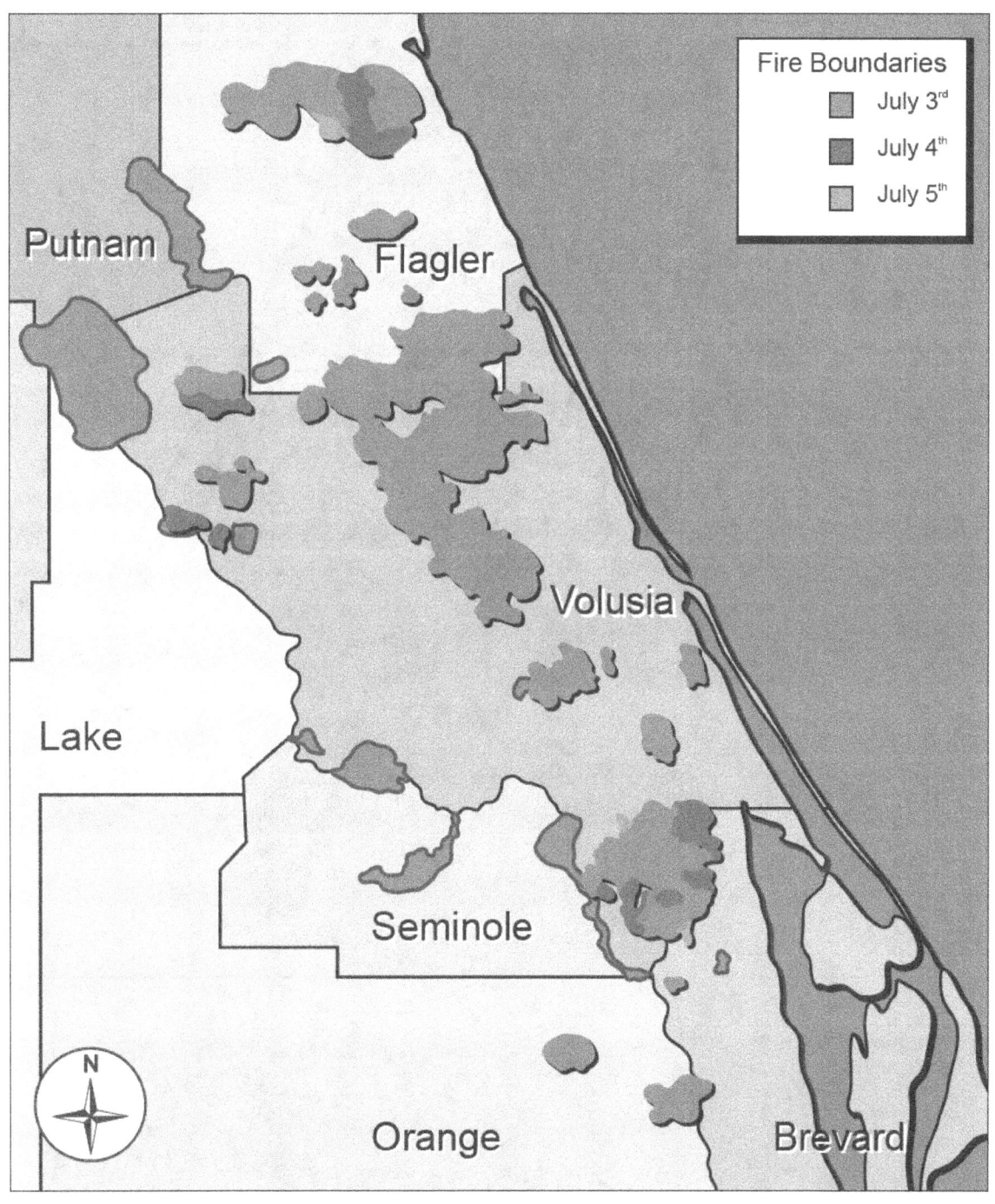

Map 1

Appendix A (continued)

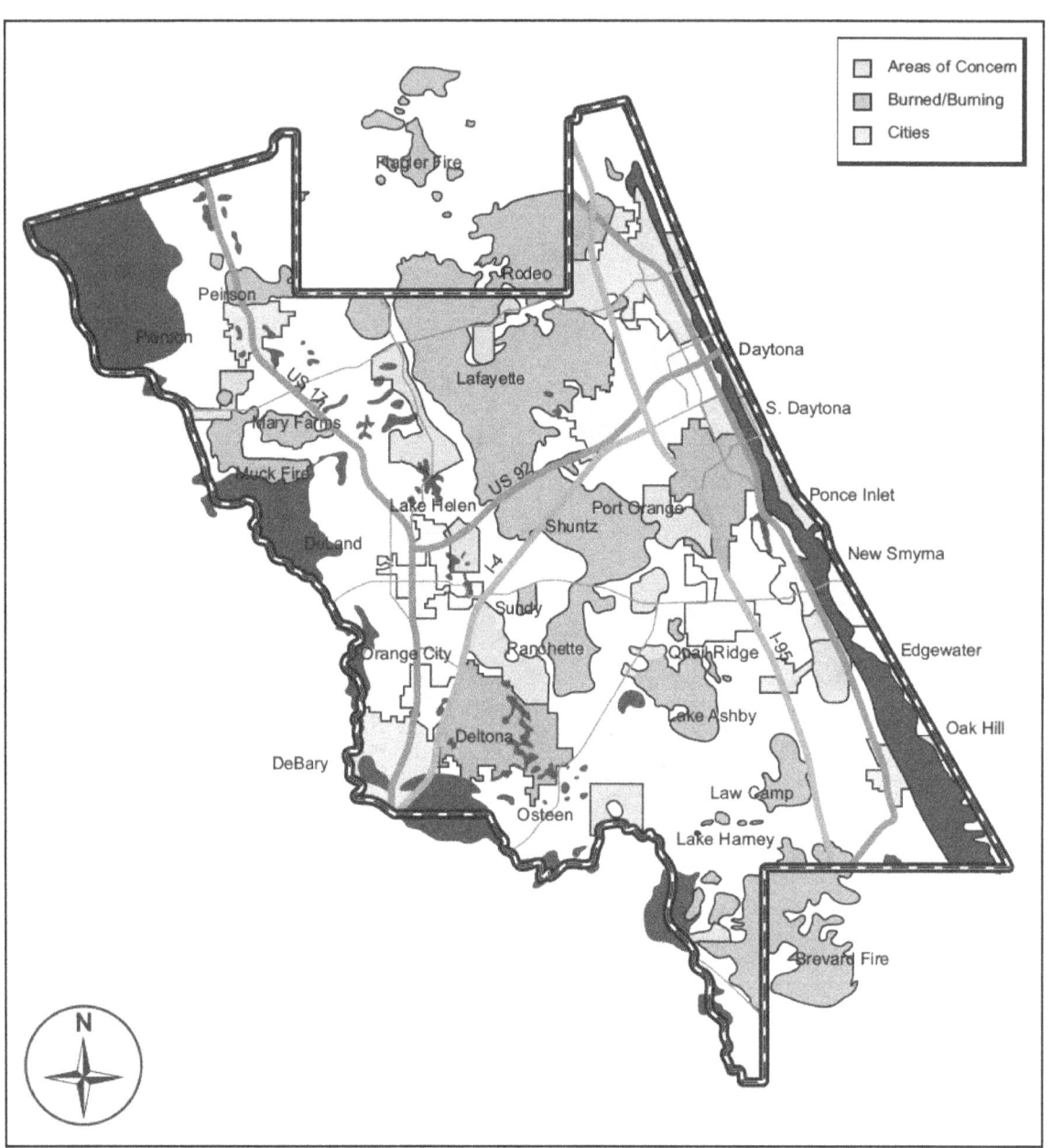

Map 2

Appendix A (continued)

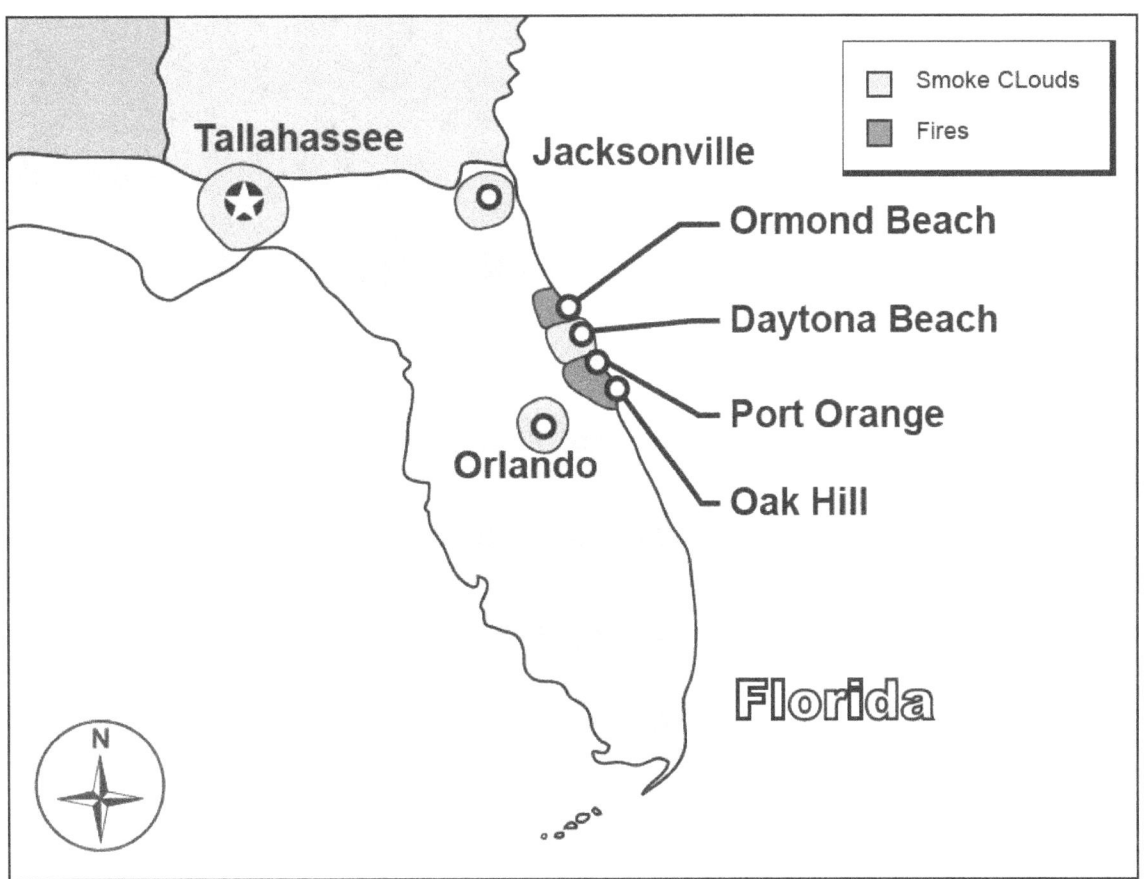

Map 3

www.ingramcontent.com/pod-product-compliance
Lightning Source LLC
Chambersburg PA
CBHW081245170526
45165CB00009B/3208